CALIFORNIA KING SNAKE AS PET

A COMPLETE GUIDE TO THEIR HABITAT, CARE, MANAGEMENT, DIET, ADAPTATIONS, BEHAVIORS AND MORE

Understanding and Husbandry

DR MORRIS HART

Copyright© 2024 **DR MORRIS HART**

All rights reserved. No part or part of this book or publication may be reproduced, stored, or transferred in any form by electronic, mechanical, recording, or other retrieval system without written permission from the publisher

Table of Contents

INTRODUCTION 5

CHAPTER 1 14

    How to Choose the Ideal California King Snake 14

CHAPTER 2 22

    Making Your California King Snake's Habitat Perfect 22

CHAPTER 3 34

    Guidelines for Your California King Snake's Nutrition and Feeding 34

CHAPTER 4 43

    Knowing the Behavior of California King Snakes 43

CHAPTER 5 53

    Tips for Managing and Introducing Your California King Snake to Others 53

CHAPTER 6 63

Maintaining the Health and Well-Being of Your California King Snake 63

CHAPTER 7 74

Considering Breeding California King Snakes 74

CHAPTER 8 87

Common Behavioral Problems with California King Snakes and Their Fixes 87

CHAPTER 9 98

Closing Thoughts: Developing a Close Relationship with Your California King Snake 98

Introduction

The fascinating California King Snake (Lampropeltis californiae) is a native of the western United States, especially California. Thanks to their eye-catching look, calm demeanor, and low maintenance needs, they have gained popularity as pets among reptile lovers across the globe. This thorough book will cover every facet of taking care of these intriguing snakes, from choosing the ideal specimen to comprehending their behavior and making sure they're healthy.

1. Natural History and Environment

The habitats that California King Snakes call home are varied and include urban areas, chaparral, grasslands, and forests. Their natural range stretches from Baja California, Mexico, to Oregon. They inhabit a range of microhabitats in these areas, including rocky hillsides and agricultural fields.

Knowing their natural past is essential to giving them the proper care when they are in captivity. We can guarantee their physical and mental well-being by trying to replicate their natural surroundings as nearly as possible.

2. Physical attributes

These snakes are well known for having exquisite looks. They usually have a smooth, slender body with shiny scales that, depending on where they live, might have different colors and patterns. Black, white, tan, and different tones of brown are common color variants that are sometimes grouped in patterned or strong bands.

With an average length of three to five feet and a female body size often greater than the male, California King Snakes have an elongated body that makes them perfect for hunting and slithering through their surroundings.

3. How to Choose the Ideal California King Snake

It's critical to choose a healthy specimen from a reliable breeder or rescue when getting a California King Snake as a pet. Seek out snakes with distinct bodily condition, smooth skin, and clear eyes. Steer clear of people who have abnormal swelling, drowsiness, or dull or sunken eyes, as these are indicators of disease or malnourishment.

Take into account the snake's disposition as well. Although most California King Snakes are calm, individual snakes might have different personalities.

Understanding the snake's temperament can be gained by handling it and seeing how it behaves.

4. Establishing the Perfect Habitat

Your California King Snake's health and happiness depend on you providing it with an appropriate habitat. It's imperative to have a large enclosure with temperature gradient, climbing branches, and safe hiding places. Provide a substrate, like aspen shavings, cypress mulch, or coconut husk, that permits burrowing and holds moisture.

Keep the temperature between the cooler 75–85°F (24–29°C) and the warmer 85–90°F (29–32°C), with a 90–95°F (32–35°C) basking area. To control temperature and avoid overheating, make use of heat sources and thermostats.

5. Guidelines for Nutrition and Feeding

As opportunistic feeders, California King Snakes take in a variety of prey items, including rodents, birds, eggs, and occasionally other reptiles. Their diet of small prey, like mice, rats, or chicks, is what keeps them healthy in captivity. Adults receive this food once every seven to ten days, while growing juveniles receive it more frequently.

To avoid regurgitation and other digestive problems, give the snake prey pieces that are about the same width as its thickest body section. To reduce tension during feeding and prevent unintentional bites, provide prey items using tongs.

6. Knowing the Behavior of California King Snakes

It's crucial to comprehend your California King Snake's behavior and cues in order to give them the best care possible. Despite their typical placid and non-aggressive nature, these snakes can become protective in response to stress or threat.

When they sense danger, they frequently coil into a defensive position, hiss, vibrate their tails, or strike. By recognizing these indicators, you may minimize stress and engage with your snake in a safe manner.

7. Managing and Socialization Pointers

It's essential to handle your California King Snake frequently in order to foster trust and socialization. As the snake grows more comfortable with your presence, start with brief, gentle handling sessions and progressively extend them.

To prevent harm and to prevent abrupt movements that could frighten or agitate the snake, support its body to the fullest extent possible. Wash your hands both before and after handling to stop the infection from spreading.

8. Maintenance of Health and Well-Being

It's critical to keep an eye on your snake's health in order to identify and address any possible problems early on. Observe their hunger, degree of activity, and general appearance closely. Respiratory symptoms, conspicuous injuries, odd feces, lethargy, and refusal to feed are some indicators of disease or distress.

It's advised to take your snake to the vet on a regular basis to make sure it stays healthy. To stop the transmission of illness, keep the cage clean and well-maintained and follow appropriate hygiene practices.

9. Breeding-Related Issues

For seasoned reptile breeders, raising California King Snakes can be a fulfilling experience. Make sure your snakes are in excellent health and have achieved sexual maturity, which is usually about 2-3 years of age, before attempting to breed them.

Place the male and female in a different breeding enclosure and keep a tight eye on their interactions. To aid with egg laying and incubation, provide suitable nesting materials and environmental circumstances.

10. Typical Behavioral Problems and Their Fixes

California King Snakes can have behavioral problems such eating refusal, excessive hiding, or hostility despite their generally calm disposition. Patience, attention to detail, and occasionally modifying husbandry techniques are needed to resolve these problems.

Providing a range of prey items, plenty of hiding places, regulating humidity and temperature, and reducing environmental stressors are common remedies.

Closing Thoughts: Developing a Close Relationship with Your California King Snake

You may establish a solid relationship with your California King Snake that is based on mutual respect and trust by being aware of its requirements and providing for them. These fascinating reptiles can live happily in captivity for many years, providing their human companions with joy and intrigue, if they receive the right care, attention, and respect for their natural behaviors.

Chapter 1

How to Choose the Ideal California King Snake

The first step in creating a happy and productive reptile-human relationship is choosing the right California King Snake as a pet. Knowing the important aspects to take into account when selecting a California King Snake will guarantee a happy experience for you and your new scaly friend, regardless of your level of experience with reptiles.

1. Study and Get Ready

Spend some time learning everything there is to know about California King Snakes before setting out on your quest to obtain one. Learn about their dietary needs, natural history, habitat requirements, and general care instructions. Knowing the fundamental requirements of

California king snakes will enable you to give your potential pet the finest care possible.

Furthermore, evaluate your personal preparedness and dedication to taking care of a snake. Take into account elements like the availability of space, the amount of time and money required, and any potential laws or ordinances in your area pertaining to the ownership of exotic pets.

2. reputable sources

After you've made the decision to bring a California King Snake into your house, you must make sure your new pet comes from a reliable rescue or breeder. Steer clear of buying snakes from dubious sources including flea markets, internet classified advertisements, or pet shops with dubious business methods.

Seek out breeders or groups who have proven to be dedicated to the well-being and moral treatment of their animals. Seek advice from other reptile lovers, take part in online communities and forums devoted to the hobby, and go to reptile expos or events where you can speak with breeders face-to-face and see their available specimens.

3. Medical Evaluation

Prioritize each snake's health and well-being when assessing possible California King Snakes. Examine every snake closely for indications of disease, trauma, or stress. A snake in good health should have glossy eyes, silky skin, and a powerful body. Snakes with mucus or secretion around the mouth or nostrils, skin lesions, or abnormalities in body structure or movement should be avoided. They should also have dull or sunken eyes.

Pay great attention to the snake's behavior to determine its general character and attitude. While some shyness or caution is natural, stay away from snakes who exhibit overly protective or aggressive behaviors as these could be signs of inadequate socialization or underlying stress.

4. Genetic Points to Consider

Take into account the genetic heritage of the California King Snake you are considering, in addition to its physical well-being. The health and genetic diversity of captive snake populations are largely determined by breeding procedures. Select breeders who place a high value on genetic diversity, ethical breeding methods, and the preservation of strong, resilient bloodlines.

Inquire with breeders about the snakes' pedigree and breeding history, as well as any known genetic characteristics or propensities for particular medical conditions. Steer clear of buying snakes from breeders who use careless or unethical breeding techniques, such as hybridization or inbreeding, as these methods can harm the health and wellbeing of the young.

5. Size and Age

There are many different ages and sizes of California King Snakes to choose from, ranging from hatchlings to adults. When choosing the age and size of your new snake, take your tastes and degree of experience into account. Hatchlings need specific care due to their unique demands and may be more delicate and sensitive to stress, but they are also more affordable and provide the chance to see their growth and development up close.

However, adult snakes may need larger enclosures and have particular food or environmental preferences. On the other hand, they may be more resilient and entrenched in their behaviors and routines. Select a snake that fits your lifestyle, expertise level, and capacity to give it the proper care for its whole life.

6. Harmony with Additional Pets

Before taking one home, think about if your other pets get along with a California King Snake. Although larger pets and humans are rarely in danger from California King Snakes, smaller animals like rodents, birds, and reptiles could be considered possible prey.

When keeping California King Snakes with other pets, take extra care, especially if the snakes are known to be aggressive toward smaller animals or have predatory instincts. To avoid mishaps or disputes, keep an eye on how your pets interact with each other and provide them with secure enclosures or separate quarters wherever possible.

7. Individual Preferences and Visual Appeal

Finally, when choosing a California King Snake, take into account your own aesthetic inclinations and preferences. California King Snakes are well-known for their eye-catching coloring and patterns, which make them excellent exhibit animals.

Pick a snake that appeals to you and fits your preferred aesthetic in terms of colors, pattern, and general appearance. Choose a snake that captures your imagination and makes you happy, regardless of whether you are drawn to traditional black and white banded specimens or uncommon mutations with striking color variations. This will increase your enjoyment of reptile care.

It's important to carefully examine a number of aspects when choosing the ideal California King Snake, including your lifestyle and tastes as well as the snake's genetic makeup and overall health. You can guarantee a happy and rewarding experience for yourself and your new scaly friend by doing extensive research, getting your snake from reliable sources, placing a high value on health and genetic diversity, and taking compatibility with other pets into consideration. Your California King Snake will flourish in its new habitat with the right care and attention, surprising you for years to come with its captivating appearance and unique habits.

Chapter 2

Making Your California King Snake's Habitat Perfect

Creating an appropriate habitat is essential to your California King Snake's health and welfare. To ensure their comfort and longevity in captivity, it is imperative to provide a setting that satisfies their physical, behavioral, and psychological demands. We will go over all the essentials of setting up the ideal habitat for your California King Snake in this in-depth tutorial, including enclosure setup, environmental enrichment, and upkeep.

1. Choice of Enclosure

Size, security, and accessibility should be your top priorities when choosing an enclosure for your California

King Snake. Your snake should have plenty of room in its enclosure to walk about, investigate, and behave in its natural manner. Larger enclosures are desirable whenever possible, with a typical rule of thumb being that one square foot of floor space should be provided for every foot of snake length.

Take into account the enclosure's design and material as well. Glass terrariums, plastic tubs, and specially constructed wooden enclosures are all workable choices, each with pros and cons. Glass terrariums have great visibility, but more ventilation might be needed to keep them from being too hot. While inexpensive and simple to maintain, plastic tubs can not be very visually appealing. Wooden enclosures that are custom-built offer insulation and customisation, but maintaining humidity levels requires careful design.

Make sure the enclosure is safe and secure, with tightly fitting doors or lids that keep your snake within and keep other animals out. To preserve air quality and avoid the accumulation of moisture and odor, make sure there is enough ventilation.

2. Selection of Substances

Selecting the appropriate substrate for your California King Snake habitat is essential for encouraging natural behaviors, preserving humidity levels, and making waste management easier. There are a number of substrates that work well with California King Snakes, each with special qualities and things to keep in mind.

Because of their great odor-controlling qualities, availability, and cost, aspen shavings are a popular option. Aspen shavings are a perfect substrate for

California King Snakes because they are non-toxic and suitable to use with reptiles.

Another good substrate alternative is cypress mulch, which has the advantages of humidity control, moisture retention, and naturalistic appearance. Cypress mulch is a strong and long-lasting substrate option since it is resistant to mold and rot.

Often referred to as "coir," coconut husk or fiber substrate is a natural and environmentally beneficial choice with superior moisture retention and odor control capabilities. The substrate made of coconut husk is perfect for your California King Snake since it helps keep the enclosure's humidity levels steady and allows the snake to dig burrows and hides.

Paper-based substrates, such as newspaper or shredded paper, are less expensive and simpler to maintain than other substrates, but they can not be as visually appealing or offer as many naturalistic advantages. Paper-based substrates can be used as a layer beneath other materials or for short-term applications.

Whichever substrate you select, make sure it's devoid of any chemicals, pesticides, or additives that could endanger your snake. To keep things clean and stop trash and bacteria from building up, replace the substrate on a regular basis.

3. Enhancement of Environment

Promoting both physical and emotional well-being requires adding realistic elements and stimulation to your California King Snake's habitat. Offer a range of hiding places, climbing frames, and items that enrich the surroundings to promote exercise, curiosity, and natural behaviors.

Using cork bark, naturally occurring rock formations, or commercially available reptile hides, add safe hiding places all over the enclosure. Your snake should feel secure in its hiding places if they are small and confined. In order to provide diversity and choice, place hiding sites in several locations throughout the cage.

Include branches and climbing frames to provide your snake opportunity to explore and exercise in the vertical space. To build stable climbing surfaces, use driftwood, branches, or PVC pipes that are firmly secured to the enclosure walls. Make sure that climbing structures are set up safely to avoid accidents or collapse.

Provide your snake with interesting and dynamic surroundings by providing imitation plants, silk foliage, and PVC pipe tunnels, among other environmental enrichment items. Try varying the colors, textures, and shapes to make an eye-catching and captivating habitat.

To provide your California King Snake with a shallow water dish to soak and drink from, think about adding one. Pick a dish that is both big enough for your snake to submerge itself completely and shallow enough to keep it from drowning. To keep the water dish hygienic and hydrated, regularly clean and refill it.

4. Lighting and Temperature

It's critical to maintain the proper temperature and lighting conditions for your California King Snake's well-being, digestion, and activity. Create a temperature gradient in the enclosure so that your snake can efficiently manage its body temperature. There should be hot basking locations and cooler retreat areas.

The temperature of the basking location should be maintained between 85 and 90°F (29 and 32°C) by placing a heat lamp or basking light over a level rock or branch inside the enclosure. Monitor basking spot temperatures with a thermometer designed specifically for reptiles, and make adjustments as necessary to keep the ideal environment.

On the cooler side of the cage, keep the ambient temperature between 75–85°F (24–29°C) and 85–90°F (29–32°C) on the warmer side. For the enclosure to be kept at the right temperature, use radiant heat panels, ceramic heat emitters, or heat pads.

Use a full-spectrum UVB light bulb or expose your California King Snake to the natural light cycle to make sure it has access to both. UVB light is necessary for reptiles' synthesis of vitamin D3 and calcium metabolism, which supports strong bone development and general wellbeing.

5. Dampness and Perspiration

Moderate humidity levels are necessary for healthy respiratory health and appropriate shedding in California King Snakes. By providing a humid hide, spraying the enclosure frequently, and utilizing materials that retain moisture, you can keep the humidity levels in the enclosure between 40 and 60 percent.

Use paper towels, coconut coir, or damp sphagnum moss in a plastic container to create a humid hiding. When your snake is shedding, place the moist hide in a warm spot of the enclosure to encourage it to be used.

Every day, softly mist the enclosure with water to raise the humidity level and keep it from drying out. To maintain ideal conditions, use a hygrometer to check humidity levels and modify the frequency of misting as necessary.

Refrain from oversaturating the substrate or letting standing water build up in the enclosure as this can encourage the spread of bacteria and cause respiratory diseases in your reptile. To stop too much moisture and humidity from building up inside the enclosure, make sure there is enough ventilation.

6. Upkeep and Cleaning

The cleanliness and well-being of your California King Snake's habitat depend on routine cleaning and upkeep. To stop waste, bacteria, and stink from building up, create a regular cleaning program that includes spot cleaning, substrate change, and enclosure disinfection.

To keep the enclosure clean and stop bacterial growth, spot clean it every day, taking care to remove any soiled substrate, uneaten food, or excrement. Substratum replacement is necessary to preserve freshness and minimize odors.

Usually every 2-4 weeks, or more frequently depending on the enclosure's state, thoroughly clean and disinfect the enclosure. Take out all of the enclosure's furniture, substrate, and environmental enrichment supplies, and give them a thorough cleaning with a disinfectant safe for reptiles.

Before putting anything back in the enclosure, make sure it's completely dry and rinsed to avoid leaving any chemical residue and to protect your snake. Keep a watchful eye out for any indications of mold, mildew, or pest infestations in the enclosure, and act quickly to remedy any problems that arise.

You may give your California King Snake the best possible living space by adhering to these recommendations and providing a well planned and maintained home. Your snake will grow and thrive in captivity and provide you with years of delight and interest if you give it the right care, pay close attention to detail, and make a commitment to meeting its behavioral and physical requirements.

Chapter 3

Guidelines for Your California King Snake's Nutrition and Feeding

For your California King Snake to be healthy, grow, and be happy, it must be fed properly. In the wild, California King Snakes are opportunistic feeders that eat a variety of prey, including rodents, birds, eggs, and occasionally other reptiles. Replicating their natural diet and using the right feeding techniques are essential to your snake's success in captivity. To assist you in giving your California King Snake the best possible nutrition, we will go over feeding schedules, choosing prey, dietary supplements, and frequent feeding problems in this extensive guide.

1. Comprehending Feeding Behavior

Understanding how California King Snakes feed is crucial before discussing particular feeding recommendations. Being omnivores, these snakes can attack their prey with lightning speed when they're hungry. They may, however, also fast occasionally, particularly during the breeding season or throughout the seasonal changes.

The majority of the time, California King Snakes eat prey that is about as wide as their thickest body portion. While juvenile snakes might need smaller prey items like fuzzy or pinky mice, adult snakes might prefer larger prey items like adult mice or small rats.

2. Feeding Schedule

For your California King Snake to remain healthy and avoid obesity or overfeeding, you must set up a regular feeding regimen. Feed adult snakes once every seven to ten days; feed growing youngsters more often, usually every five to seven days.

To keep your snake at a healthy weight and bodily condition, keep an eye on its condition and change the frequency of feedings as necessary. Refrain from feeding your snake too often as this can result in obesity and related health issues. On the other hand, malnutrition and stunted growth may arise from underfeeding.

California King Snakes may go into a state of brumation, or slowed metabolic activity, during the winter, resulting in decreased activity and appetite. During this time, reduce the number of times you feed your child, serve smaller meals, or skip feeding entirely.

3. Selection of Prey

Choosing the right prey items is essential to feeding your California King Snake a healthy, well-balanced diet. To replicate their natural diet and supply vital nutrients, provide a range of prey items. Typical prey items include quail eggs, chicks, mice, and rats.

Select prey items according to the size and age of your snake. Choose prey items for adult snakes that are approximately one to 1.5 times the diameter of the snake's thickest body section. Select smaller, more easily ingested and digested prey items for juveniles.

Make sure the prey comes from reliable sources and is devoid of pesticides, parasites, and other toxins that could injure your snake. Before giving frozen prey to your snake, make sure it is completely thawed to avoid digestive problems and to guarantee that the nutrients are properly absorbed.

4. Feeding Procedure

Offering your California King Snake prey items can be done in a few different ways: live feeding, pre-killed prey, and frozen-thawed prey. Select the approach that is most effective for both you and your snake, as each has pros and downsides.

By providing your snake with live prey, you enable it to engage in its natural hunting habits. Even though some snakes may prefer to eat on live prey, doing so entails some hazards, like the snake becoming harmed by protective prey or contracting parasites or diseases from the prey.

Pre-killed prey is prepared for your snake by first putting it to sleep. By using this strategy, you can feed your snake in a safer manner and lessen the chance of injury. Utilize compassionate euthanasia techniques like cervical dislocation or CO2 gas to guarantee the prey item dies quickly and painlessly.

Prey that has been frozen and thawed provides a practical and secure substitute for live or pre-killed prey. You can find frozen prey items easily from pet supply stores and online vendors, which you can keep in your freezer until you need them. Before giving frozen prey to your snake, make sure it is completely thawed to avoid digestive problems and to guarantee that the nutrients are properly absorbed.

5. dietary supplements

You may need to provide your California King Snake with dietary supplements in addition to whole prey items to make sure it gets all the nutrients it needs. In reptiles, calcium and vitamin D3 are especially crucial for bone health and metabolic processes. Give your snake some prey by dusting it with a calcium supplement that contains vitamin D3.

To avoid oversupplementation and related health risks, supplements should be taken rarely and moderately. Regarding dosage and supplementation frequency, heed manufacturer recommendations and keep an eye on the general health and wellbeing of your snake.

6. Typical Feeding Problems

You could still run into frequent feeding problems with your California King Snake, even with your best efforts. Obesity, regurgitation, and food rejection are a few examples of these problems. In order to resolve these problems, one must be patient, vigilant, and occasionally adapt feeding procedures.

There are many different reasons why someone could refuse to eat, such as stress, illness, or environmental influences. Make sure your snake's habitat is properly set up, offer hiding places and suitable environmental enrichment, and keep an eye out for any symptoms of disease or suffering.

Regurgitation may happen if your snake is handled too soon after eating, consumes an excessive amount of food, or encounters other stressful situations. To reduce the chance of regurgitation, give your snake plenty of time to digest its food without being disturbed. Also, refrain from handling it for at least 24 to 48 hours after feeding.

Overfeeding or providing larger-than-comfortable prey items for your snake to eat can lead to obesity. To keep your snake at a healthy weight, keep a close eye on its physical condition and make any necessary adjustments to the amount and frequency of feedings.

The two most important parts of taking care of your California King Snake are feeding and nutrition. Your snake can get the nutrition it needs to flourish in captivity if you are aware of its natural feeding habits, set up a regular feeding schedule, choose suitable prey, and provide it necessary dietary supplements. Keep an eye on your snake's physical health, make any necessary adjustments to its diet, and get advice from a veterinarian who specializes in reptile care if you have any questions or concerns about feeding. Your California King Snake will make a long, healthy, and happy friend if you provide them the care and feeding they require.

Chapter 4

Knowing the Behavior of California King Snakes

To provide your California King Snake the best care and build a solid relationship with your pet, you must be aware of its behavior. California King Snakes display a diverse array of behaviors that provide important insights into their individual personalities and overall well-being. These activities vary from their natural instincts and communication cues to their social interactions and environmental preferences. We will delve into the intriguing realm of California King Snake behavior in this thorough book, which covers subjects including handling, social behavior, thermoregulation, hunting and feeding, and typical behavioral problems.

1. In the Wild, Natural Behavior

Native to the western United States, California King Snakes can be found in a variety of environments such as woodlands, grasslands, chaparral, and urban areas. They are mostly nocturnal hunters in the wild, using their strong constriction and excellent sense of smell to catch and eat a wide range of food, such as eggs, rodents, birds, and other reptiles.

California King Snakes may hide during the day in dense vegetation, underground tunnels, or rock crevices in order to evade predators and maintain body temperature. Because they are skilled climbers, they may take use of trees, bushes, and other vertical formations for hiding, hunting, and sunbathing.

Most of the year, California King Snakes live alone, but during the breeding season, they may gather in large groups. While females emit pheromones to entice possible mates, men compete for mating opportunities through combat rituals.

2. Behavior Related to Thermoregulation

The California King Snake, like all other reptiles, is an ectothermic animal, meaning that it gets its body warmth from outside sources. For snakes to maintain proper digestion, activity levels, and physiological processes, thermoregulation behavior is essential.

Give your captive California King Snake a temperature gradient in its enclosure, with a warmer basking region and a colder retreat section. Snakes will thermoregulate, alternating between these temperatures to keep their bodies at the ideal range, usually between 85 and 90 degrees Fahrenheit (29 and 32 degrees Celsius).

Your snake's health and well-being can be greatly inferred from its thermoregulation behavior. A healthy, responsive snake will exhibit alert, active behavior and frequent movement between temperature zones; on the other hand, extended periods of inactivity or stationary behavior may indicate underlying medical conditions or environmental stressors.

3. Behavior in Hunting and Feeding

The varied food of California King Snakes includes rodents, birds, eggs, and occasionally other reptiles. They are opportunistic feeders. They happily take in a wide range of prey items in captivity, including as mice, rats, chicks, and quail eggs.

California King Snakes use their keen sense of smell to find their prey when they are hunting. They can actively pursue and apprehend escaping prey by using their speed and agility, or they can use ambush tactics to surprise and overwhelm their target.

California King Snakes use constriction to immobilize and subdue their victim after they have it and then eat it whole. They might also rip and swallow larger prey items with their strong jaws and recurved teeth.

Snakes can differ in how they behave when they are feeding; some may respond enthusiastically, while others may be more circumspect or picky. Make sure your snake eats its prey effectively and safely by keeping a close eye on its feeding habits.

4. Social Conduct

Although California King Snakes are mainly solitary creatures, they can display social behaviors when they get together in communal hibernation locations or during the breeding season. While females emit pheromones to entice possible mates, males compete for mating opportunities and establish dominance through combat rituals.

California King Snakes usually engage in courtship and mating actions in the spring or early summer. Males can participate in complex displays of courting that involve physical conflict with competitor males and rubbing their chin glands on things in the surrounding area.

The average nesting season for female California King Snakes is late spring or early summer. Depending on their age, size, and surroundings, the clutch size can range from 6 to 20 eggs. In order to guarantee the proper development of the eggs, females may display maternal behaviors such as watching over the nest site and controlling the humidity and temperature inside.

5. Managing and Introducing

One of the most crucial aspects of socializing and trust-building is handling your California King Snake. As your snake gets used to you, start with brief, gentle handling sessions and progressively extend the time and frequency.

Strike your snake with confidence and composure, moving slowly and deliberately so as not to frighten or upset it. When holding your snake, provide complete support to its body to avoid harm and reduce tension.

Frequent handling will lessen your snake's anxiety during veterinarian examinations and other handling procedures and help desensitize it to human contact. On the other hand, if your snake exhibits indications of discomfort or distress, you should always respect its boundaries and cues and never force or confine it.

6. Typical Behavioral Problems

California King Snakes can have behavioral problems such eating refusal, excessive hiding, or hostility despite their generally calm disposition. Patience, attention to detail, and occasionally modifying husbandry techniques are needed to resolve these problems.

There are many different reasons why someone could refuse to eat, such as stress, illness, or environmental influences. Make sure your snake's habitat is properly set up, offer hiding places and suitable environmental enrichment, and keep an eye out for any symptoms of disease or suffering.

The presence of external stressors in the snake's surroundings, insufficient environmental circumstances, or stress itself may be indicated by excessive hiding behavior. Make sure the lighting, temperature, humidity, and cage configuration are all suitable for your snake.

Although it is uncommon, aggression in California King Snakes can happen when they sense stress or threats. To reduce stress on the snake, eliminate abrupt movements or loud noises that could frighten or agitate it. Instead, offer safe hiding places and enriching surroundings.

It's essential to comprehend the behavior of California King Snakes in order to give your pet the best care and foster a close relationship. You may learn a lot about each person's unique personality, state of health, and communication patterns by paying attention to their social interactions, natural instincts, and communication signs. Your California King Snake will flourish in captivity and add fascinating behaviors and a captivating presence to your life if you are patient, meticulous, and dedicated to addressing their behavioral needs.

Chapter 5

Tips for Managing and Introducing Your California King Snake to Others

Building trust, encouraging socialization, and fortifying your relationship with your reptile friend all depend on how you handle your California King Snake. Handling sessions may be fun for both you and your snake if you use the right methods, are patient, and show respect for its own preferences. We will go over socialization techniques, typical handling problems, and safe and effective handling guidelines in this extensive guide to help you build a fulfilling connection with your California King Snake.

1. Building Confidence

You must build a relationship of trust and familiarity with your California King Snake before beginning handling sessions. Spend some time in the vicinity of the snake's enclosure so that it gets used to your smell and presence. Move gently and speak softly to prevent frightening or upsetting the snake.

Providing food during handling sessions can facilitate the association of your presence with favorable outcomes. Present your snake with prey items using feeding tongs so it can feed off of your hand. Your snake will eventually learn to correlate your presence with positive reinforcement and identify you as a food source.

It's best to avoid handling your snake just after feeding to reduce the chance of regurgitation and digestive problems. Before beginning handling sessions, give your snake enough time to finish eating without interruption.

2. Correct Methodology

It's critical to handle your California King Snake with appropriate method to protect your safety and the comfort of the snake. Strike your snake with confidence and composure, moving slowly and deliberately so as not to frighten or upset it.

When holding your snake, provide complete support to its body to avoid harm and reduce tension. Support the snake's weight evenly across its body as you gently lift it up from underneath. Retaining or gripping the snake firmly can make it uncomfortable and may make it react defensively.

Handle your snake in a safe, contained space devoid of escape routes or possible threats. To avoid mishaps or injuries, keep your snake away from open doors, windows, and other pets.

3. Slow Introductory

As your snake grows more comfortable with your company, progressively increase the length and frequency of handling sessions. Start with brief, gentle sessions. First, stay within the enclosure and let the snake investigate your hand or arm.

Lift the snake out of the enclosure with caution and provide stable support for its body once it seems at ease with your presence. As the snake becomes more at ease and habituated to handling, start with quick handling sessions that last no longer than a few minutes and progressively extend them.

During handling sessions, pay attention to the behavior and body language of your snake. Stress or discomfort might be indicated by quick breathing, hissing, protective posture, or a sudden desire to get away. Put your snake back in its enclosure and try again later if it exhibits these symptoms.

4. Enhancement of Environment

During handling sessions, use environmental enrichment to stimulate your California King Snake's mind and promote its natural activities. Provide the snake with opportunity to explore; for example, let it climb on branches or investigate various surfaces and textures.

During handling sessions, give your snake safe havens and hiding places to go to if it gets overstimulated or agitated. To create a dynamic and engaging setting, include elements that enhance the surrounding environment, such as PVC pipe tunnels, silk foliage, and artificial plants.

Keep things that provide environmental enrichment fresh and avoid habituation by rotating them frequently. Try a variety of textures, shapes, and scents to make your handling sessions engaging and stimulating for your snake.

5. Encouragement that is positive

During handling sessions, use positive reinforcement tactics to encourage your California King Snake to behave calmly and cooperatively. As a means of support and encouragement, give little food prizes, soft pats, or verbal praise.

You may make handling sessions more enjoyable by providing food incentives or having interactive play right before or after handling. Your snake may become more calm and cooperative in subsequent sessions as it learns to associate handling with positive feelings.

During handling sessions, stay away from utilizing punishment or negative reinforcement approaches since these can make your snake fearful and mistrustful. Rather, concentrate on developing a cooperative and respectful partnership that is founded on mutual respect.

6. Observing Limits

During handling sessions, pay attention to your California King Snake's cues and boundaries. Keep an eye out for telltale symptoms of tension or discomfort, such hissing, protective posture, or trying to run away. If your snake exhibits these symptoms, carefully put it back in its cage and give it another go at a later time.

Steer clear of overhandling or prolonged handling of your snake as this can cause tension and exhaustion. Avoid handling your snake right before or right after feeding or shedding, and give it plenty of time to recuperate between handling sessions.

Seek advice from an expert reptile keeper or reptile behavior specialist if your snake routinely displays protective or aggressive tendencies during handling sessions. They can provide knowledge and methods for dealing with behavioral problems and fostering a healthy bond with your snake.

7. Managing Children vs. Adults

Because California King Snakes vary in size, strength, and temperament, handling juvenile and adult specimens may call for various techniques. Because they are usually more apprehensive and cautious than adults, juveniles may need more time and consideration while handling them.

When working with young snakes, begin with brief, gentle sessions and progressively extend the length and frequency as the snake gets used to the handling. Take additional care when providing support for their fragile bodies, and steer clear of abrupt movements that could frighten or upset the snake.

During handling sessions, adult California King Snakes tend to be more at ease and self-assured, although they may yet display protective behaviors if they sense stress or danger. Adult snakes should be approached with poise and patience, moving slowly and deliberately to prevent setting off defensive reactions.

During handling sessions, always put your snake's comfort and safety first, regardless of age. Be ready to modify your handling methods and strategy in response to the unique temperament and behavior of your snake.

Taking good care of your California King Snake and developing a close relationship with your reptile friend needs handling and socialization. You may create positive and fulfilling experiences for both you and your snake by building trust, utilizing appropriate handling skills, and paying attention to your snake's indications and boundaries. Your California King Snake can become a treasured and respected member of your family with time, effort, and a dedication to developing a trustworthy bond.

Chapter 6

Maintaining the Health and Well-Being of Your California King Snake

Maintaining your California King Snake's health and well-being is crucial to enhancing its lifespan and standard of living. You may give your reptile pet the best care possible by following good husbandry procedures, keeping an eye out for any symptoms of disease or injury, and getting timely veterinary attention when necessary. The essential elements of maintaining the health and well-being of California King Snakes will be covered in detail in this extensive book, covering habitat management, feeding and nutrition, sanitation and hygiene, preventive healthcare practices, and common health problems.

1. Management of Habitat

Sustaining the health and welfare of your California King Snake depends on keeping its habitat tidy and well-planned. Maintaining your snake's habitat regularly keeps it free of bacteria, parasites, and waste, as well as providing a secure and comfortable living space.

Select an appropriate substrate for your snake's enclosure, such as fiber from coconut husks, aspen shavings, or cypress mulch. Substratum replacement is necessary to keep things clean and stop bacteria and garbage from accumulating.

Temperature and Humidity: Use digital thermometers and hygrometers to keep an eye on the enclosure's temperature and humidity conditions. Create a temperature gradient with the warmer side being 85–90°F (29–32°C) and the colder side being 75–85°F (24–29°C). Keep humidity levels between 40 and 60 percent to promote healthy respiration and appropriate shedding.

Environmental Enrichment: To promote natural behaviors and mental stimulation, provide hiding places, climbing frames, and other environmental enrichment objects. To keep enrichment materials fresh and avoid habituation, rotate them frequently.

Cleaning and upkeep: Give the enclosure regular cleaning and upkeep, which should include replacing the substrate, disinfecting the surfaces and furniture, and spot cleaning. Provide fresh, clean water every day and clean water dishes on a regular basis.

2. Nutrition and Feeding

For California King Snakes to maintain development, energy, and general health, they must receive a proper diet. To make sure your snake eats a healthy, balanced diet, abide by these rules:

Choose Your Prey: Provide a range of prey items, such as rats, mice, chicks, and quail eggs. Select prey items according to the size and age of your snake.

Feeding Schedule: Growling juveniles should be fed more frequently, usually every 5-7 days, whereas adult snakes should be fed once every 7–10 days. Depending on your snake's bodily condition, metabolism, and activity level, you can adjust the frequency of feedings as necessary.

Dietary Supplements: To enhance bone health and metabolic function, dust prey items with a calcium supplement containing vitamin D3 prior to delivering them to your snake. To avoid supplementing too much, use vitamins sparingly and in moderation.

Monitoring: Keep a close eye on your snake's physical state and feeding behavior. In order to maintain a healthy weight and physical condition, modify feeding habits as necessary.

3. tidiness and hygiene

It's crucial to keep your California King Snake clean and hygienic in order to stop disease from spreading and to support its overall health. Maintain the cleanliness of your snake's habitat by adhering to these hygiene practices:

Hand Washing: To stop the spread of bacteria and parasites, properly wash your hands with soap and water both before and after touching your snake or cleaning its enclosure.

Cleaning and disinfection of the enclosure should be done on a regular basis. This includes clearing away any filthy substrate, washing all of the furniture and surfaces, and sanitizing the water dishes. Adhere to the manufacturer's instructions for correct usage and dilution when using disinfectants meant for reptiles.

Substrate Replacement: To keep things clean and stop bacteria and waste from accumulating, replace the substrate on a regular basis. Make use of non-toxic and safe substrate materials for reptiles.

Water Quality: Every day, fill a shallow dish with clean, fresh water that is big enough for your snake to soak in and drink from. To stop bacteria from contaminating water dishes and algae from growing, clean them frequently.

4. Preventive Healthcare Practices

By putting preventative healthcare practices into practice, you can lower your California King Snake's risk of sickness and preserve its general health and wellbeing. Observe these recommendations for preventive healthcare:

Frequent Veterinary Check-ups: To keep an eye on your snake's health and identify any possible problems early, schedule routine wellness checks with a veterinarian who is knowledgeable about reptiles. Consult your veterinarian about the proper immunization schedule and preventive care options.

Control of Parasites: Keep an eye out for any indications of parasites in your snake, such as diarrhea, weight loss, or strange behavior. To create a parasite management strategy that is specific to your snake's requirements, speak with your veterinarian.

Keep new reptiles you add to your collection quarantined to stop the spread of infectious diseases. For at least 30 days, keep newly acquired snakes apart from more experienced ones, and keep a watchful eye out for any symptoms of disease.

Environmental Management: To limit stress and lower the chance of disease transmission, maintain ideal habitat conditions, including temperature, humidity, and cleanliness. Your snake's physical and mental well-being can be supported by feeding it a balanced food and providing suitable environmental stimulation.

5. Typical Health Concerns

Your California King Snake may occasionally still have health problems even with your best efforts. Learn about common health problems in snakes and how to identify the symptoms and signs:

Respiratory infections: Lethargy, open-mouth breathing, wheezing, and nasal discharge are among symptoms. If you suspect a respiratory infection in your snake, get veterinary attention as soon as possible.

Mites and Ticks: Snakes are susceptible to infestations by external parasites like mites and ticks, which can irritate and discomfort them. Regularly check your snake for indications of external parasites, and discuss treatment options with your veterinarian.

Shedding Issues: Poor shedding in snakes can be caused by environmental variables such as low humidity levels. To make shedding easier, make sure the hide is humid and the humidity levels are appropriate. If your snake has trouble shedding or retains shed, get veterinary care.

Digestive Problems: An inappropriate food, poor feeding techniques, or unfavorable environmental circumstances can all lead to constipation, regurgitation, and other digestive problems. Keep a watchful eye on the digestive system and feeding response of your snake, and contact your veterinarian if any problems develop.

A mix of good husbandry techniques, preventative healthcare measures, and timely veterinarian care when necessary are needed to maintain the health and wellness of your California King Snake. You can make sure your snake has a long, healthy, and happy life in captivity by giving it a clean, well-designed home, a nutritious feed that is balanced, and regular monitoring for signs of disease or injury. If you have any worries about the health or wellbeing of your snake, get competent veterinarian care right away. Remain watchful, proactive, and sensitive to your snake's requirements. Your California King Snake will flourish and continue to captivate and delight you for many years to come with the right care and attention.

Chapter 7

Considering Breeding California King Snakes

For reptile enthusiasts, raising California King Snakes can be a fulfilling endeavor, but it does take careful planning, preparation, and understanding of the reproductive biology and behavioral patterns of the species. Successful breeding necessitates meticulous attention to detail and a dedication to giving the best care possible for both the adult snakes and their progeny, from matching breeding couples and setting up suitable breeding settings to tending to gravid females and incubating eggs. We will go over all of the important factors that need to be taken into account while breeding California King Snakes in this thorough guide, including choosing a breeding pair, scheduling the breeding season, setting up the habitat, wooing and mating behaviors, egg incubation, caring for the newborn, and typical breeding problems.

1. Breeding Pair Selection

An essential first step in effective breeding is choosing compatible breeding couples. Select genetically varied, healthy snakes with desirable characteristics including pattern, color, and temperament. Steer clear of breeding related persons to reduce the likelihood of genetic defects and health problems in the progeny.

Take into account several characteristics like size, age, and past reproductive success when choosing breeding pairings. Compared to immature or subadult snakes, mature, sexually mature snakes have a higher chance of successfully reproducing. Select snakes that are sexually mature and at least two to three years old. Male and female snakes should normally be somewhat smaller than two feet in length.

Before introducing possible breeding pairs for reproduction, watch how well they get along and interact with one another. Select partners who exhibit compatibility and mutual interest, and keep an eye out for symptoms of violence, stress, or apathy.

2. Timing of the Breeding Season

Breeding season for California King Snakes usually begins in the spring or early summer, when the days get longer and the temperature rises. The timing of the breeding season can vary based on a number of variables, including climate, location, and individual reproductive cycles.

For the best time to breed, keep an eye on environmental cues and factors like temperature, humidity, and photoperiod. Increase the number of sunshine hours gradually and maintain the right temperature and humidity levels to replicate the circumstances found during natural breeding.

Breeding pairs should be introduced to one another gradually to give them time to get used to one another's company and form social relationships. Keep a watchful eye on their behavior for indications of courtship and readiness for mating, such as circling, rubbing, and marking of the chin gland.

3. Configuration of Habitat

In order to encourage productive breeding practices and the development of eggs, it is imperative to create the optimal breeding habitat for California King Snakes.

Assemble a roomy, well-ventilated container with the right humidity, temperature, and environmental enhancements.

Size of Enclosure: Select a roomy enclosure that can comfortably house both breeding couples. Provide a variety of hiding places, climbing frames, and items that enhance the surrounding environment to encourage natural behaviors and lower stress levels.

Temperature and Humidity: To replicate natural breeding conditions, keep the enclosure's temperature and humidity levels at a certain level. Create a temperature gradient with the warmer side being 85–90°F (29–32°C) and the colder side being 75–85°F (24–29°C). Keep humidity levels between 40 and 60 percent to promote healthy shedding and reproduction.

Nesting Substrate: Provide gravid females with an appropriate nesting substrate to deposit their eggs on, such as sphagnum moss, vermiculite, or a combination of peat moss and sand. To keep the ideal circumstances for egg incubation and to stop mold growth, make sure the nesting substrate is damp but not soggy.

4. Mating and Courtship Behaviors

To ensure successful reproduction, California King Snakes perform intricate courtship rituals and mating activities. Males may use aggressive actions to show off their dominance and draw in potential partners, including as chin rubbing, tail shaking, and body wrapping.

Keep a keen eye out for cues from breeding couples, such as circling, rubbing, and marking of the chin gland, that indicate they are ready for mating. Males and females should be introduced gradually to give them time to form social relationships and exhibit courting behaviors.

When the male and female are ready to mate, bring them inside after the wooing behaviors have been seen. To guarantee successful copulation and reduce the possibility of hostility or harm between breeding partners, keep a watchful eye on mating sessions.

5. Severe Female Assistance

Giving pregnant female California King Snakes the nourishment, habitat, and nesting requirements they require is especially important. Give gravid females enough food and fluids to encourage the development of their eggs and to keep them healthy and happy overall.

Gravid females should be closely observed for any indications of stress, discomfort, or pregnancy-related problems. To aid in egg laying and incubation, make sure they have access to an appropriate nesting spot with wet nesting substrate.

Give gravid females space and solitude as they build their nests and lay their eggs. Minimize disruptions and deal with them as little as possible to lessen anxiety and encourage the proper development and incubation of eggs.

6. Incubation of Eggs

To guarantee the health and viability of the progeny, gravid females must be given ideal conditions for egg incubation after they have set their eggs. To ensure the best possible humidity retention and egg growth, move the eggs to a different incubation container that is filled with wet vermiculite or perlite.

To replicate the conditions of a natural nest, keep the incubation container's temperature and humidity levels constant. For ideal egg incubation, maintain a consistent temperature range of 78–82°F (25–28°C) and humidity levels of 75–85%.

Keep a watchful eye on the development of the eggs, looking for indications of mold, dehydration, or egg viability. To keep the environment ideal for viable eggs and avoid contamination, remove any infertile or rotten eggs as soon as possible.

7. Infant Care

After the eggs hatch, give neonatal California King Snakes the proper care and handling to guarantee their health and welfare. Move the hatchlings to a different rearing habitat that has the right amount of humidity, light, and air.

After they have shed their first skin, give neonatal snakes suitable sized prey items, like fuzzy mice or newborn pinky mice. Keep a constant eye on their growth and feeding response, and modify the prey's size and feeding schedule as necessary to promote optimal development.

To encourage natural behaviors and lessen stress in young snakes, provide them lots of hiding places and enrichment items for their surroundings. Provide fresh, clean water in a shallow dish and keep a constant eye on your pet's hydration, especially while they are eating and shedding.

8. Typical Breeding Problems

When breeding California King Snakes, frequent breeding problems might still arise despite your best efforts. These problems could be egg binding, infertility, aberrant development, or problems with the health of the newborn.

Environmental stressors, insufficient compatibility between breeding partners, or underlying medical conditions in one or both breeding partners can all contribute to infertility. If infertility problems persist, get a consultation with a reptile veterinarian and keep a close eye out for symptoms of courtship and mating readiness in breeding pairs.

If a gravid female is unable to deposit her eggs for any reason—such as insufficient nesting substrate, dehydration, or physical obstruction—egg binding, also known as dystocia, may result. If veterinary intervention and supportive care are required to aid in egg laying and avoid problems, provide them.

Neonate snakes may exhibit developmental anomalies as a result of incorrect incubation techniques, environmental variables, or genetics. When problems emerge, get veterinarian attention and keep a watchful eye out for any indications of developmental abnormalities in the hatchlings.

Erroneous husbandry techniques or environmental stressors can lead to health problems in newborns such as malnourishment, dehydration, and respiratory infections. To maintain the health and wellbeing of newborn snakes, give them the proper attention and seek veterinarian assistance when necessary.

For those who love reptiles, breeding California King Snakes can be a difficult but rewarding task. You may encourage successful reproduction and generate healthy progeny by choosing breeding partners that get along well, creating ideal breeding conditions, and providing proper care for gravid females and newborn adult snakes. Throughout the breeding process, be watchful, proactive, and sensitive to your snakes' requirements. If you run into any problems, get expert veterinarian attention. You may experience the excitement of breeding California King Snakes and help to preserve and conserve this amazing species with the right planning, preparation, and care.

Chapter 8

Common Behavioral Problems with California King Snakes and Their Fixes

Promoting the wellbeing of California King Snakes and maintaining a positive relationship between the snake and its owner need an understanding of and response to typical behavioral problems in these animals. California King Snake habits can range from aggressive and stress-related behaviors to feeding issues and habitat preferences; these behaviors call for attention and management. We will examine the most typical behavioral problems that snake owners face in this extensive guide, along with workable strategies to successfully treat and reduce these problems.

1. Behaviors Related to Stress

California King Snakes can exhibit a variety of stress-related behaviors, such as protective posture, hiding, refusing to eat, and excessive activity. Promoting a healthy and happy snake requires determining and treating the root sources of stress.

Solution: Reduce environmental stresses for the snake by giving it enough hiding places, keeping its temperature and humidity levels constant, and causing as little disruption as possible when it is being fed, handled, and cleaned. Make sure the enclosure is situated in a peaceful region away from busy streets and noisy surroundings.

2. hostility

Despite their typical gentle and non-aggressive nature, California King Snakes can become defensive when they sense stress or danger. Hissing, hitting, or biting are examples of aggressive behaviors that occur when someone feels threatened or disturbed.

Solution: To prevent inciting defensive reactions, approach your snake with composure and slow, deliberate movements. To lessen stress and lower the likelihood of violence, give the snake time and space to become used to its surroundings and handle it with gentleness and respect.

3. Issues with Feeding

Feeding difficulties, including selective feeding, regurgitation, and unwillingness to eat, might be brought on by stress, poor husbandry practices, or underlying medical conditions. It's critical to take quick action when there are feeding issues with your snake to guarantee it gets enough food and to preserve its general health and wellbeing.

Solution: To guarantee ideal feeding conditions, assess the enclosure and husbandry techniques used for the snake. To promote nutritional variety, offer prey items that are adequately proportioned and alter the diet. To promote good feeding behavior, carefully monitor feeding reactions and make necessary adjustments to feeding methods.

4. Preferences for habitats

The ideal habitat for California King Snakes varies depending on the temperature, humidity, and level of environmental enrichment. Inadequate provision of appropriate habitat conditions may cause stress, disease, and behavioral problems in confined snakes.

Solution: Learn about the natural environment of California King Snakes and try to mimic it as much as you can in a captivity. To assist the physical and emotional well-being of the snake, provide it with a temperature gradient, hiding places, climbing structures, and environmental enrichment.

5. Managing Problems

Overhand or improper handling methods can stress and discomfort California King Snakes, making them defensive or withdrawing. A healthy relationship between the snake and its owner depends on you being aware of your snake's preferences and honoring its boundaries when handling it.

Solution: Treat your snake with care and sensitivity, supporting its body and reducing stress with appropriate handling techniques. As the snake gets used to handling, start with brief, gentle handling sessions and progressively increase the length and frequency. During handling sessions, pay attention to the snake's indications and body language and respect its boundaries.

6. Environmental Stressors

Environmental stressors including low humidity, illumination, temperature, or substrate can cause stress and behavioral problems in California King Snakes. Promoting the health and well-being of the snake requires limiting environmental stressors and creating a suitable habitat.

Solution: Use digital thermometers and hygrometers to keep an eye on the enclosure's temperature and humidity conditions. To replicate natural illumination and create a temperature gradient, adjust the lighting and heating appropriately. Make sure the substrate you choose for the snake's burrow is secure, non-toxic, and cozy for it to explore.

7. Behavior in Territories

California King Snakes have been known to act in a territorial manner, particularly when living in a group or near other snakes. Territorial aggression can take the form of aggressive behavior toward alleged intruders, rivalry for resources, or dominance displays.

Solution: To reduce aggressiveness and territorial issues, give every snake its own enclosure. If the snakes aren't compatible breeding partners, don't house more than one together, and keep a close eye out for any indications of stress or violence in their behavior. Give each snake plenty of hiding places and enrichment in the surroundings to lessen rivalry for resources and increase their sense of security.

8. Disinterest and Passivity

Stress, sluggishness, and behavioral problems can arise from providing California King Snakes with insufficient environmental stimulation or enrichment, which can cause boredom and inactivity. It is imperative to offer opportunities for both cerebral and physical stimulation in order to enhance the general health and well-being of the snake.

Solution: To promote natural behaviors and mental stimulation, provide environmental enrichment items such hiding places, climbing structures, and interactive toys. To keep enrichment materials fresh and avoid habituation, rotate them frequently. Give the snake chances to explore and interact with its surroundings to keep it interested and active.

9. Excessive Management

California King Snakes that are handled excessively may get agitated and uncomfortable, which may result in defensive actions, retreat, or decreased hunger. Promoting the snake's wellbeing and lowering stress levels requires balancing handling sessions with times of rest and quiet.

Solution: Restrict handling to brief, gentle sessions; do not handle right before or right after feeding, shedding, or other stressful occasions. Respect the snake's desire for privacy and rest by paying attention to its behavior and signs. Give the snake plenty of places to hide and retreat so it can unwind away from people.

10. Not Enough Socialization

California King Snakes that have not been socialized or handled properly may exhibit fear, protective tendencies, or hostility, particularly if they have not been around people since they are young. Building confidence and trust in captive snakes requires appropriate handling and socialization methods.

Solution: Begin interacting with your snake early on to help it become less fearful or aggressive. Treat the snake with decency and gentleness, supporting its body and reducing tension with appropriate handling techniques. As the snake gets used to human contact, gradually extend the length and frequency of handling sessions. During handling sessions, exercise patience, consistency, and respect for the snake's boundaries and cues.

In order to effectively manage typical behavioral problems in California King Snakes, one must be patient, vigilant, and proactive in learning about the requirements and preferences of the snake. You can encourage a happy, healthy snake and a good rapport between the owner and the pet by figuring out the root causes of behavioral problems and putting suitable remedies in place. If you see any persistent or serious behavioral problems in your snake, get professional veterinary care. Otherwise, be watchful, proactive, and sensitive to your snake's behavior and overall health. Your California King Snake will flourish and continue to captivate and delight you for many years to come with the right care and attention.

Chapter 9

Closing Thoughts: Developing a Close Relationship with Your California King Snake

Patience, understanding, and devotion are necessary to forge a close bond with your California King Snake, which is a pleasant and fulfilling experience. Building a strong bond with your snake improves both of your lives and increases your admiration for these amazing reptiles, from giving them the best care and habitat to encouraging trust and respect. We will consider the experience of interacting and caring for your California King Snake in this concluding guide, and we will conclude by discussing the significance of preserving this unique link.

1. Understanding the Special Features of California King Snakes

California King Snakes are fascinating and charming companions because of their distinctive set of characteristics and activities. California King Snakes are wonderful pets and companions because of everything about them, including their sleek and colorful appearance, gentle disposition, and curious curiosity. Spend some time observing and appreciating the unique characteristics and tastes of your snake, and cherish the unique relationship you two have.

2. Giving the Best Possible Care and Husbandry

Building a solid relationship with your California King Snake and encouraging longevity and vigor depend on you taking care of its health and well-being. An environment that is secure and comfortable for your snake to live in can be created by giving it the best possible care and husbandry, which includes a wholesome diet, appropriate housing, and frequent veterinary examinations. To promote your snake's general health and well-being, pay attention to their behavioral and physical cues and alter their care as necessary.

3. Building Mutual Respect and Trust

Developing mutual respect and trust is the cornerstone of a solid relationship with your California King Snake. When interacting with your snake, maintain a composed and self-assured demeanor, employing tactful handling methods while honoring their personal space and preferences. As you establish a relationship built on mutual respect and trust, take the time to learn about your snake's preferences and dislikes. Also, be patient and understanding.

4. Recognizing and Meeting Behavioral Requirements

It's crucial to comprehend the behavioral requirements and preferences of your California King Snake in order to foster a happy and satisfying relationship. Pay attention to their nonverbal signs and body language, and give them opportunities for socializing, interactive play, and environmental enrichment to stimulate their minds and bodies. To ensure the wellbeing of your snake, take prompt and proactive measures to address any behavioral abnormalities and seek advice from knowledgeable reptile keepers or veterinarians as needed.

5. Accepting the Pleasures of Owning a Snake

Having a California King Snake as a pet is an amazing and fulfilling experience that will enrich your life with happiness, curiosity, and company. Spend quality time with your snake, watch their activities, and educate yourself on their biology and natural history to fully enjoy the benefits of owning a snake. Take great satisfaction in giving your snake the best care possible and developing a close relationship based on mutual respect, trust, and understanding.

6. Promoting Growth and Lifelong Learning

Taking care of a California King Snake presents chances for personal development and enrichment as it is a lifelong learning and growth experience. Continue to be inquisitive and receptive, looking for fresh knowledge and perspectives on the behavior and upkeep of snakes. To benefit the larger reptile-keeping community, make connections with other hobbyists and share your expertise and experiences.

7. Treasured Times Spent Together

You will make lifelong memories with your California King Snake, and each moment you spend with it is valuable and significant. Whether you're feeding your snake your favorite dish, watching them explore their habitat, or just spending time together as they warm up in their enclosure, you should treasure the moments you have with your snake. The special tie that exists between you and your snake is created during these moments of companionship and connection.

8. Considering the Trip

Take satisfaction in the advancements you've made and the relationship you've established with your California King Snake as you consider your journey of caring for and interacting with it. Honor your accomplishments, no matter how modest, and keep in mind the obstacles you overcame on the path. Your unwavering devotion, affection, and love for your snake have improved their quality of life and profoundly enhanced your own.

Developing a close relationship with your California King Snake is an exciting adventure of learning, development, and enrichment for both of you. You can develop a deep and meaningful relationship with your snake that will enrich your life and bring happiness by giving it the best care and husbandry possible, encouraging mutual respect and trust, and embracing the joys of snake ownership. Treasure the times you spend with your snake, and keep growing and strengthening your relationship as you both go through ups and downs in life. Your California King Snake will flourish and provide you with great joy and company for many years to come if you have the patience, understanding, and dedication to providing the best possible care for your pet.

www.ingramcontent.com/pod-product-compliance
Lightning Source LLC
Chambersburg PA
CBHW050115230526
45470CB00004B/1845